Timothy Lightoler

The Gentleman and Farmer's Architect

A New Work, Containing a Great Variety of Useful and Genteel Designs

Timothy Lightoler

The Gentleman and Farmer's Architect
A New Work, Containing a Great Variety of Useful and Genteel Designs

ISBN/EAN: 9783337102982

Printed in Europe, USA, Canada, Australia, Japan

Cover: Foto ©berggeist007 / pixelio.de

More available books at **www.hansebooks.com**

THE

GENTLEMAN and FARMER's

ARCHITECT.

A NEW WORK.

Containing a great VARIETY of

USEFUL and GENTEEL DESIGNS.

BEING CORRECT

PLANS and ELEVATIONS

O F

PARSONAGE and FARM HOUSES,

Lodges for Parks, Pinery, Peach, Hot and Green Houses,

WITH THE

FIRE-WALL, TAN-PIT, &c. particularly described.

DUTCH, and other Barns, Cow-Houses, Stables, Sheepcots, Huts, Facades ;

With all other Offices appertaining to a well-regulated Farm ;

Their Situations rendered convenient, and Aspects agreeable.

WITH

SCALES and TABLES of REFERENCE,

Describing the several Parts, with their just Dimensions and Use.

DESIGNED and DRAWN

BY

T. LIGHTOLER, Architect.

And well Engraved on Twenty-five Folio Copper-Plates.

L O N D O N :

Printed for ROBERT SAYER, Map and Print-Seller, at the Golden-Buck, near Searjeant's-Inn,
Fleet-Street.

MDCCLXII.

Parsonage House

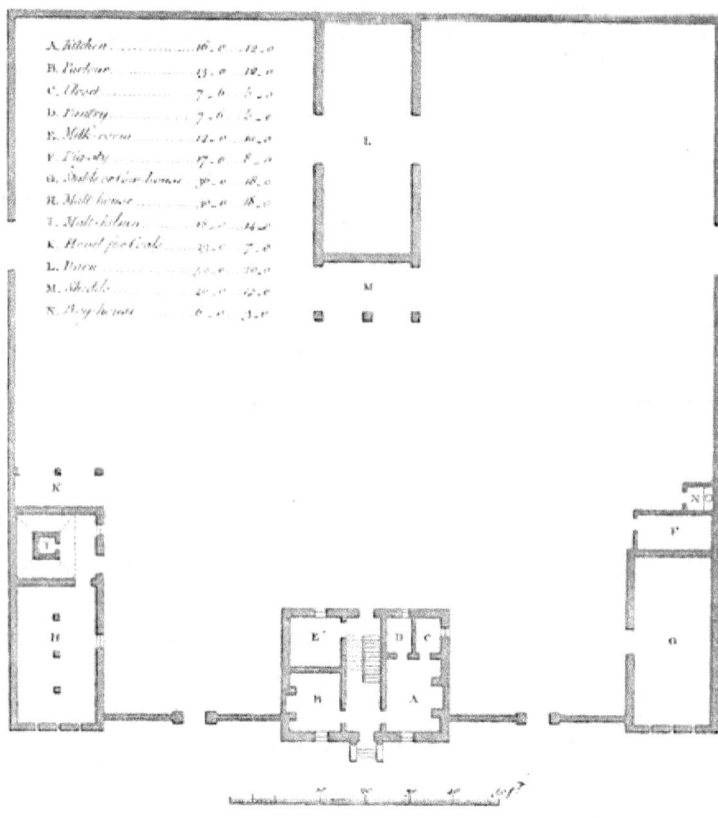

A.	Kitchen	16.0 ..... 12.0
B.	Parlour	13.0 ..... 12.0
C.	Closet	7.6 ..... 5.0
D.	Pantry	7.6 ..... 5.0
E.	Milk room	12.0 ..... 10.0
F.	Laundry	17.0 ..... 8.0
G.	Stable or cow house	30.0 ..... 18.0
H.	Malt house	30.0 ..... 18.0
I.	Malt kilns	18.0 ..... 14.0
K.	House for Cask	22.0 ..... 7.0
L.	Barn	40.0 ..... 20.0
M.	Stable	20.0 ..... 12.0
N.	Dog house	6.0 ..... 5.0

The Plan and Elevation of Parsonage or Farm House.

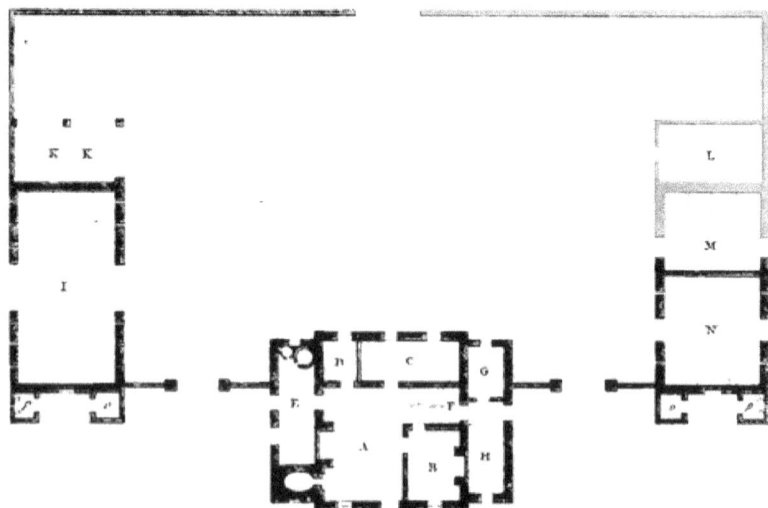

A.	The Kitchen	21 6 15 6	G.	Cellar	10 0 7 0	N. Stable	20 0 19 0
B.	Parlour	14 0 10 0	H.	Coalhouse &c.	17 6 7 0	O.	
C.	Milk Room	10 6 8 0	I.	Barn	36 0 19 0	P. Hoghouse	
D.	Pantry	6 0 7 0	K.	Hovel			
E.	Wash and Brewhouse	22 0 9 0	L.	Hoggesty			
F.	Stairs and Passage	10 0 7 0	M.	Stable	19 0 15 0		

Small Parsonage House

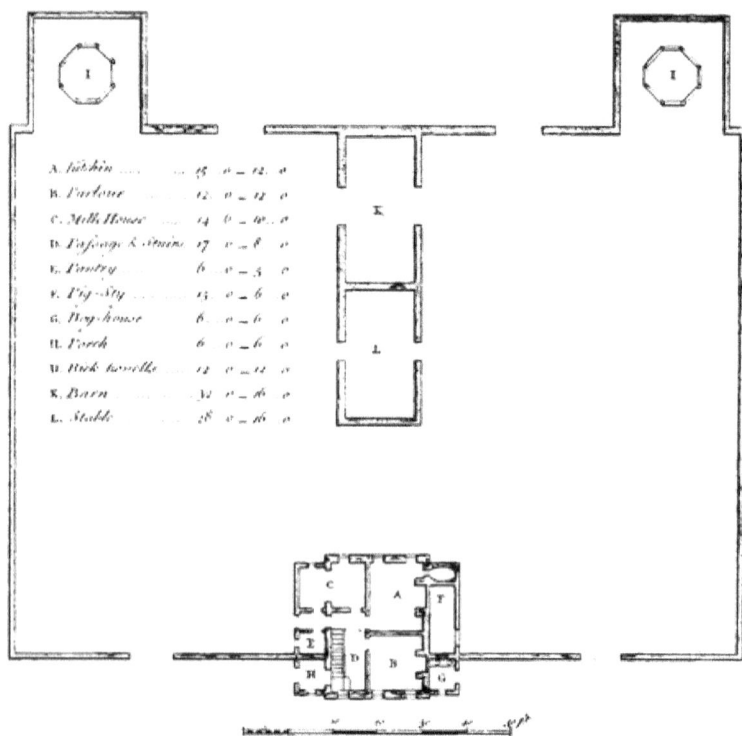

A. Kitchen .... 15 0 — 12 0
B. Parlour .... 12 0 — 12 0
C. Milk House .... 14 0 — 10 0
D. Passage & Stairs 17 0 — 8 0
E. Pantry .... 6 0 — 5 0
F. Pig-Sty .... 13 0 — 6 0
G. Wash-house .... 6 0 — 6 0
H. Porch .... 6 0 — 6 0
I. Back house .... 12 0 — 12 0
K. Barn .... 32 0 — 16 0
L. Stable .... 18 0 — 16 0

Two small Farm Houses.

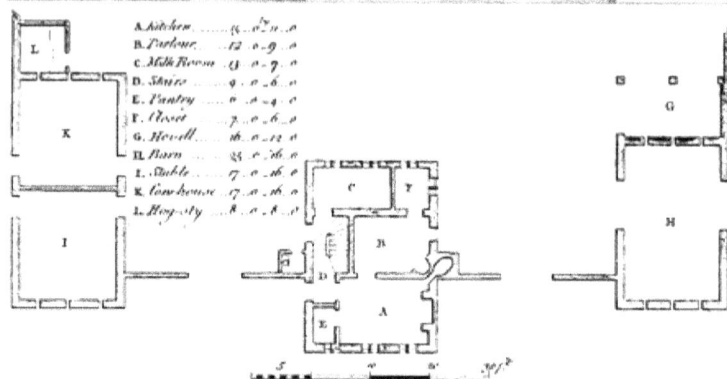

5

Back front to Barn

Elevation of House

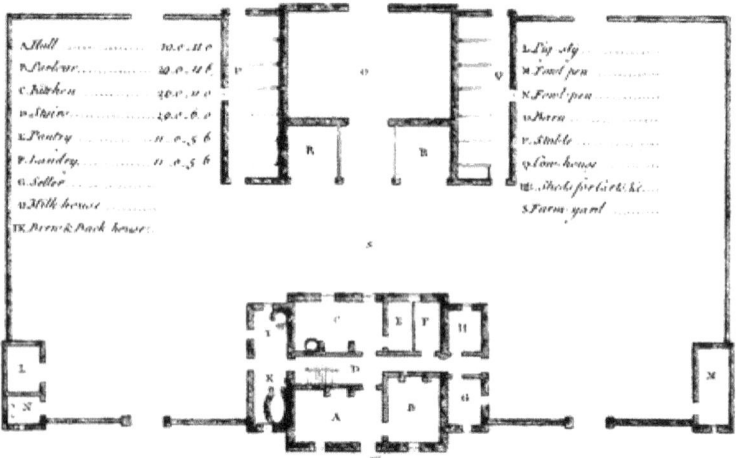

A Hall	10.0.11.0
B Parlour	20.0.11.6
C Kitchen	20.0.11.0
D Stair	20.0.6.0
E Pantry	11.0.5.6
F Laundry	11.0.5.6
G Seller	
D Milk house	
IK Brew & Back house	

L Pig sty	
M Fowl pen	
N Fowl pen	
O Barn	
P Stable	
Q Cow house	
R Sheds for Carts &c	
S Farm yard	

6

The Plan & Elevation of a Farm House & Offices

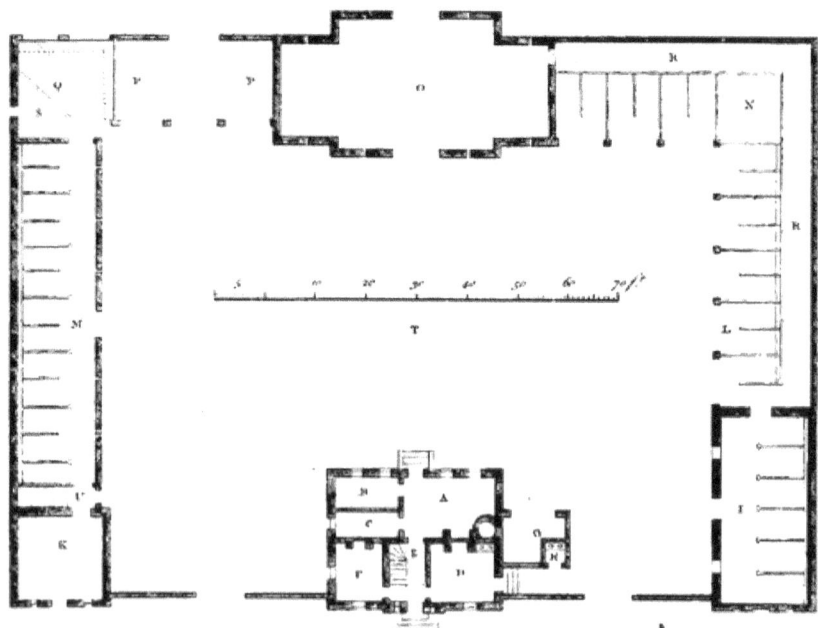

A.	The common Kitchin	18	9	12	0	H. Bog house		P.P. Hovells for Waggons	
B.	Milk room	12	3	6	0	I. Stable	35 0 17 0	O. Hogsty	
C.	Pantry	12	3	5	8	K. Granary	17 0 16 0	R.R.	
D.	Back kitchen	14	0	11	0	L. Ox Shedds		S. Calf house	
E.	Passage & Staircase	11	0	7	6	M. Cow house	65 0 16 0	T. Barn Yard	
F.	Parlour	11	0	8	8½	N.		V.	
G.	Coal Yard					O. Barn	53 0 35 0		

.

A. *Barn* .... 43 . 0 . 21 . 0
B. *Stable &c* ... 43 . 6 . 48 . 0
C. *Ox Stalls* ... 9 . 0 . 9 . 0
D. *Hay Housing* . 9 . 0 . 9 . 0
E. *Passage* .
F. *Farm-house* .. 3? . 0 . 21 . 0
G. *Shed & Granary*
   over it .....

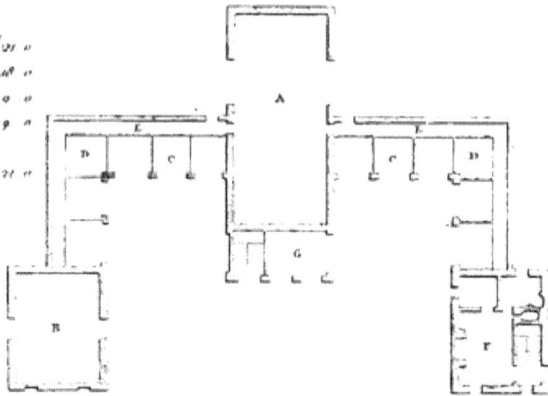

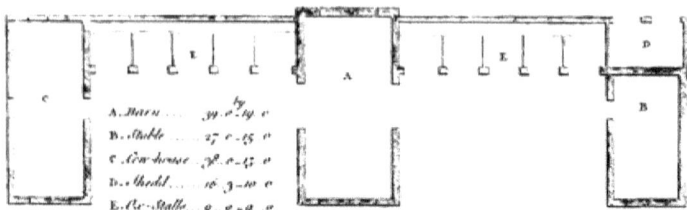

A. *Barn* ..... 34 . 0 . 14 . 0
B. *Stable* .... 27 . 0 . 15 . 0
C. *Cow-house* . 38 . 0 . 17 . 0
D. *Shed* ...... 16 . 3 . 10 . 0
E. *Ox Stalls* .. 9 . 0 . 9 . 0

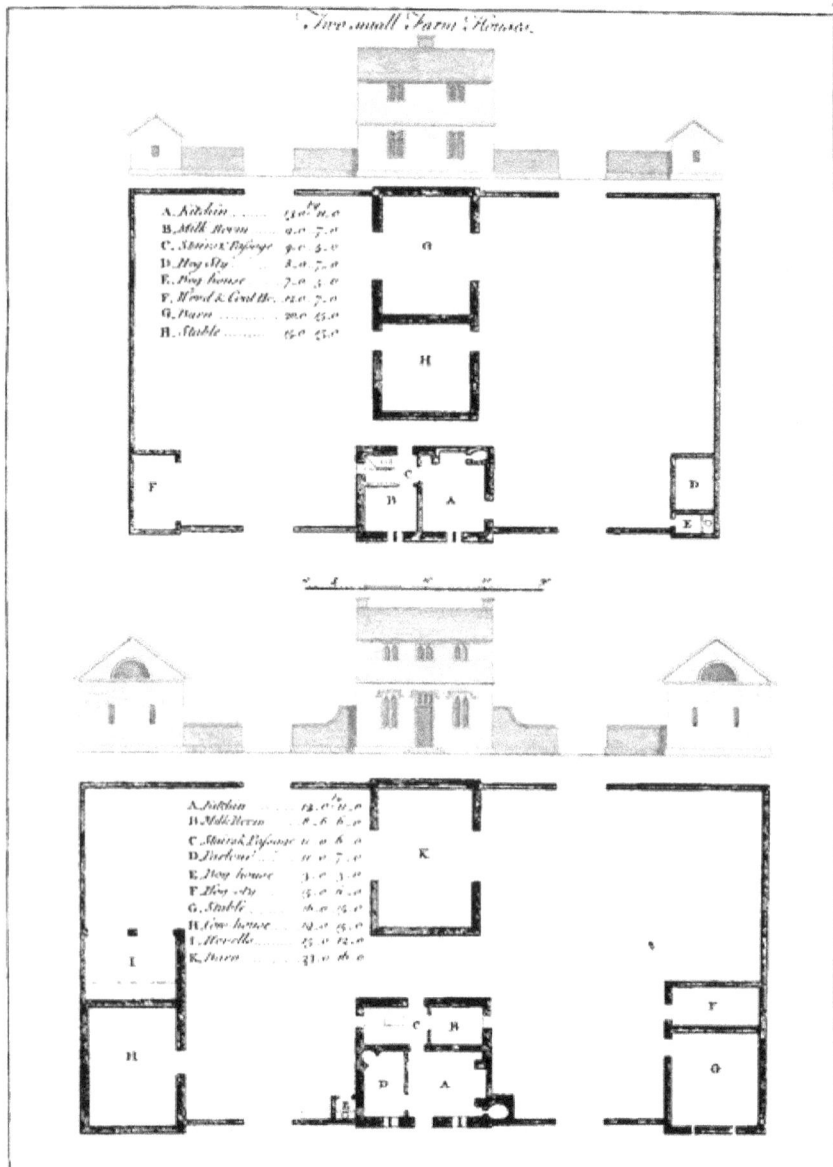

Two small Farm Houses

A. Kitchin ....
B. Milk Room ....
C. Shower Passage ....
D. Hog Sty ....
E. Dog house ....
F. Wood & Coal He. ....
G. Barn ....
H. Stable ....

A. Kitchin ....
B. Milk Room ....
C. Shower Passage ....
D. Parlour ....
E. Dog house ....
F. Hog Sty ....
G. Stable ....
H. Cow house ....
I. Novells ....
K. Barn ....

Plan & Elevation for a Dairy Farm.

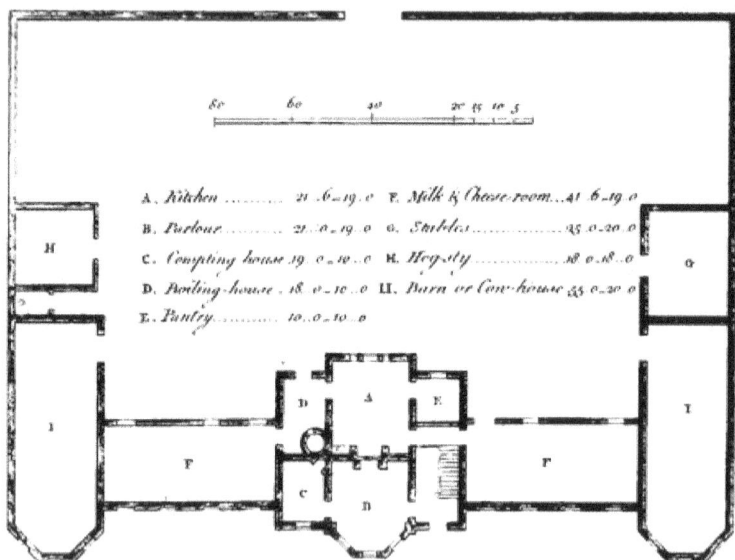

A. Kitchen .... 21.6—19.0   F. Milk & Cheese-room...41.6—19.0
B. Parlour .... 21.0—19.0   G. Stables .... 25.0—20.0
C. Compting house 19.0—10.0   H. Hogsty .... 18.0—18.0
D. Boiling-house 18.0—10.0   II. Barn or Cow-house 55.0—20.0
E. Pantry .... 10.0—10.0

Farm House?

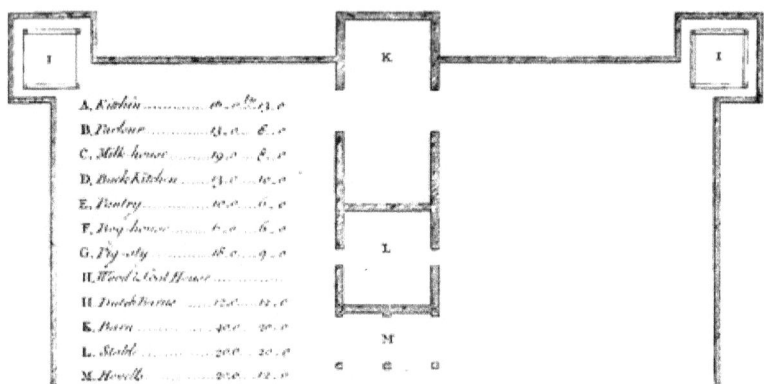

A. *Kitchin* .... 10. 0 by 13. 0
B. *Parlour* .... 13. 0 .... 6. 0
C. *Milk house* ...... 19. 0 .... 8. 0
D. *Back Kitchen* .... 13. 0 .... 10. 0
E. *Pantry* .... 10. 0 .... 6. 0
F. *Hog house* .... 6. 0 .... 6. 0
G. *Pig sty* .... 18. 0 .... 9. 0
H. *Wood & Coal House* ....
H. *Dutch barns* ..... 120. 0 .... 12. 0
K. *Barn* .... 40. 0 .... 20. 0
L. *Stable* .... 200. 0 .... 20. 0
M. *Hovells* .... 250. 0 .... 12. 0

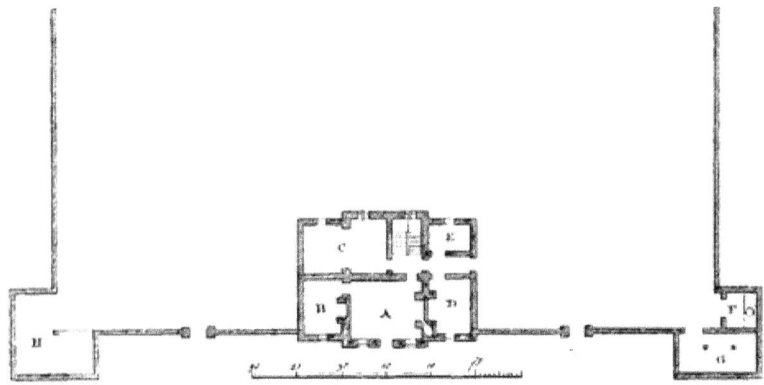

Plan & Elevation of a Lodge-house adapted for any part where a Dairy is kept.

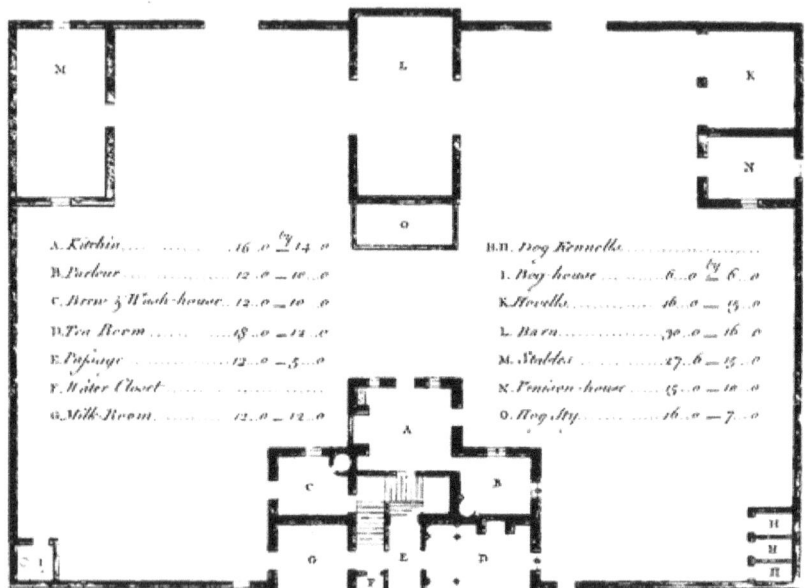

A. Kitchin	16.0 by 14.0
B. Parlour	12.0 — 10.0
C. Brew & Wash-house	12.0 — 10.0
D. Tea Room	18.0 — 14.0
E. Passage	12.0 — 5.0
F. Water Closet	
G. Milk Room	12.0 — 12.0
H.H. Dog Kennells	
I. Hog house	6.0 by 6.0
K. Hovells	16.0 — 13.0
L. Barn	30.0 — 16.0
M. Stables	17.6 — 13.0
N. Venison house	15.0 — 10.0
O. Hog Sty	16.0 — 7.0

0   5   10   20   30   40   50   60

Plan & Elevation of a Hott house ?

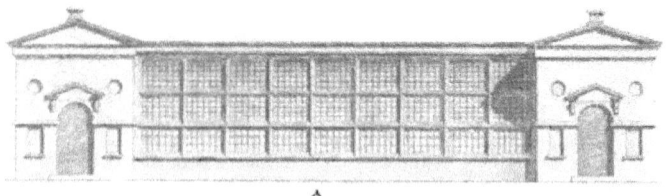

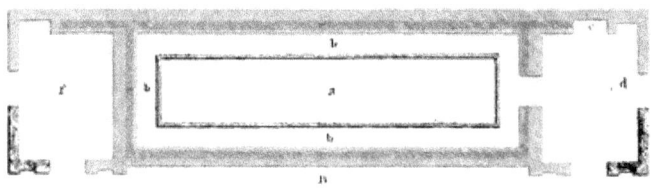

A.

B

A. *Elevation .* ——————
B. *Plan .* ——————
a. *Tan pitt* ——————
bb. *Wall round Tan-pitt .* ——————
c. *Fire place by that heat. the Walls.*
d. *Steps to descend to the fire-place ?*

ee . *The flue discoverd by the deper shade n*
flue is bent below the floor at the does . }
f . *The Green-house.*
g. *The Roller for Canvas taken up by means* }
of a Wheel & Line ? . . . . }

Section of the Hott-house.

A. Section lengthways.
B. Section across the Wing building with the Steps to the Fire-place and door into the Hott-house.
C. Section across the middle of the Hott-house.

aa. Tan-pitt.
bb. Fire-place to heat the Flews.
cc. The Flews mark'd by dotted Lines &c.
dd. The Green and Fowel-houses.
ee. The Roller in front & end.

Plan and Elevation of a Fire wall.

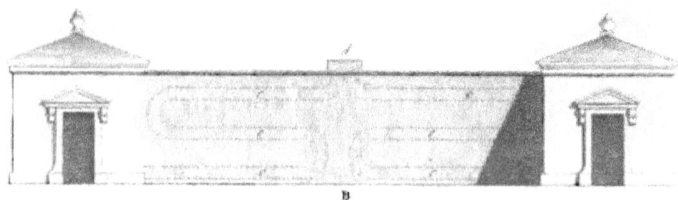

A

B

A . The Plan.

B . The Elevation

c.c. The Houses for Fewell and Heating
the Flues

a.a. The Fire places

b.b. The Steps to descend to the Fire places
3 f.t 6 Inches below the Ground

cccc.The Flews in the Plan & Elevation
marked by Prickt lines

d. The Chimney

Design of a Peach and Vine Wall.

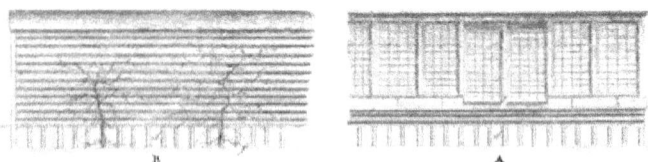

A. The Elevation of part of the Extent —

B. shews the Battning or Boards nail'd on Slips of Wood which are fasten'd to the Wall, C. The Section a cross —

d. d. d. is a Bed of Tan and cover'd with a Strata of Earth on which may be planted Strawberrys Mellons &c —

e As a Walk fill'd with Earth for the Roots of the Trees to shoot in, and wherever they are planted there must be an Arch turn'd in the Wall for the Roots to spread as shewn by the dotted — Line at f. Note the Windows need not be lower'd far but by a few Pins as shewn at g.

The Plan & Elevation of a Lodge or Keepers House fit for a Park. Fore cast, &c.

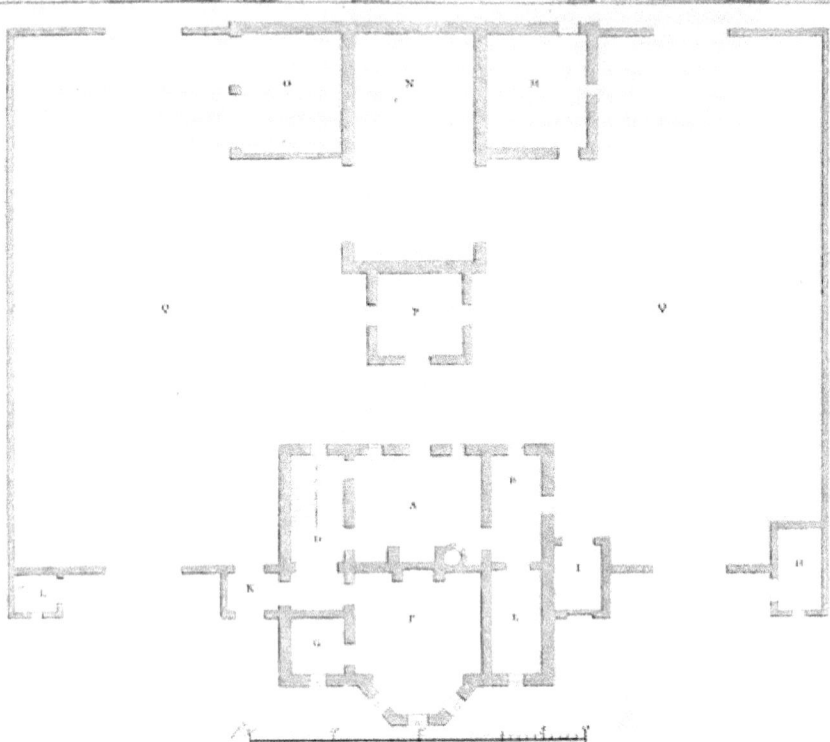

A. Kitchen	13.3 0 0	R. Coal house	0 0 0 0
B. Dairy	11 0 7 0	K. Porch	0 0 4 0
C. Pantry under the Stair	7 0 3 0	L. Hog-house	6 0 4 0
D. Passage		M. Stable	13 0 12 0
E. Milk house	12 0 7 0	N. Barn	20 0 13 0
F. Parlour or Tea Room	15 3 10 0	O. Herd	13 0 12 0
G. Closet	7 0 7 0	P. Venison house	10 0 10 0
H. Hog sty	0 0 3 0	Q. Barn-yard	90 0 60 0

Plan & Elevation of a Lodge House & Farme, in the Chinese Taste.

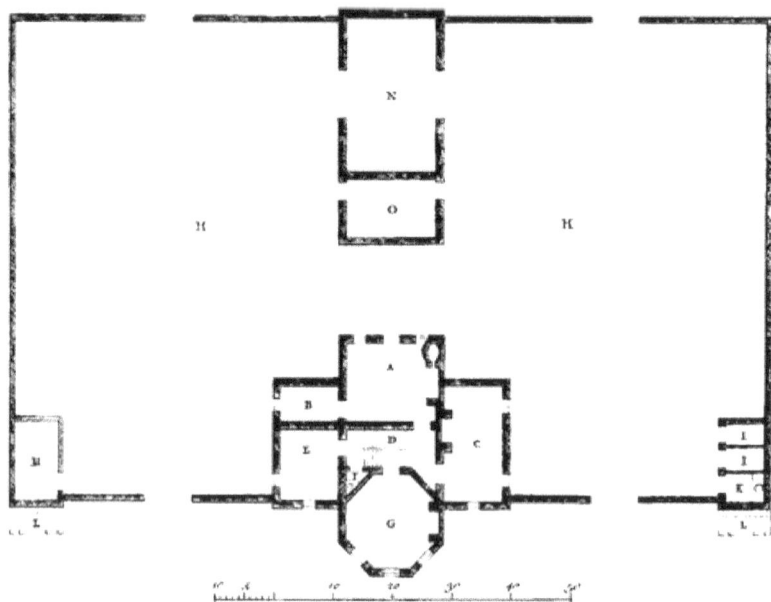

A	Kitchen	15	0	11	0	
B	Pantry	9	0	5	6	
C	Parlour	19	0	10	0	
D	Staircase	15	0	6	0	
E	Milk Room	10	4	10	0	
F	Water Closet					
G	Tea Room	15	0	15	0	

HH	Barn and Rick-yard					by
II	Dog-kennel	6	6	7	6	
K	Boy-house	4	6	4	6	
L	Alcove Seats	9	0	4	0	
M	Hogg-sty	12	0	6	6	
N	Barne	25	0	15	0	
O	Stable	15	0	10	0	

Out Offices.

A. Round yard & Kennels...... 66..0 ⚹ 37..0	G. Spaniel yard and Kennels ..15..0 ⚹ 20..0
B. Feeding yard.............. 17..0 — 26..0	H. Boiling house................14..0 — 14..0
C. Passage 9 feet 6 Inches wide...............	I. Hall........................23..0 — 12..0
D. Four Horse Stable........23..6 — 16..0	K. Parlour for the Keeper.......14..0 — 11..0
E. Ox or Cow yard & Stable..51..0 — 51..0	L. Pig house.
F.F. Passage 6 feet wide	

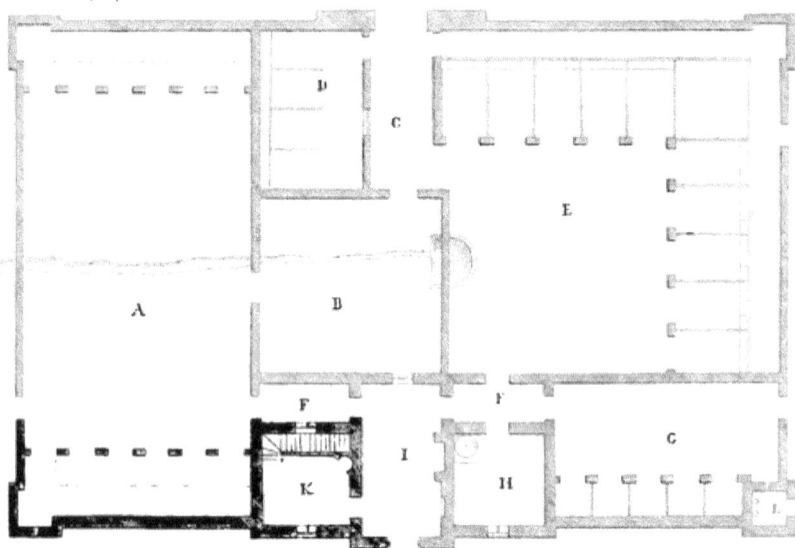

_Profile to Plate 18.

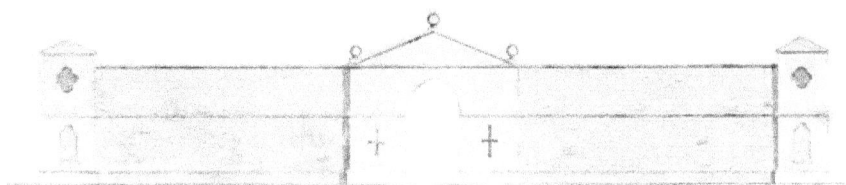

_Back Front to Plate 18.

Section to Plate 18.

T. Lightoler Inv.ᵗ   S. Miller sc.

Kitchen Dairy, &c.

A. Kitchen . . . . 17 6 by 13 6
B. Scullery and Bakehouse . . . . . . 15 6 — 6 0
C. Larder . . . . 12 0 — 6 0
D. Dairy . . . . 12 0 — 6 0
E. Passages, 5 feet 6 inches wide . . . .
F. Laundry . . . . 12 0 — 15 0
G. Washhouse . . . . 15 6 — 13 0
H. Brewhouse . . . . 15 6 — 13 0

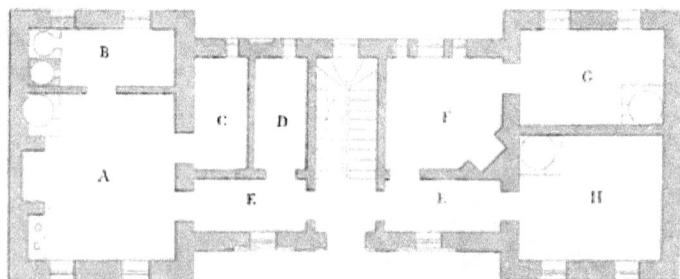

Coach houses, Stables &c.

A. Passage to Coach yard .... 37 .. 6½ .. 11 .. 0
B. Stairs and Passage 6 feet wide ....
C.D. Coachman and Grooms Rooms .... 12 .. 0 .. 11 .. 6
E.F. Stables .... 30 .. 0 .. 18 .. 0
G. Stairs 6 feet wide ....
H.I. Coach Houses .... 12 .. 0 .. 11 .. 0

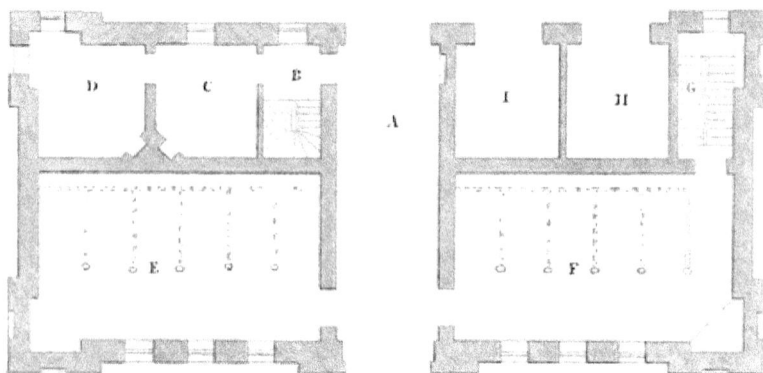

P. Baslutsder Invt.   F. Waller sc.

.

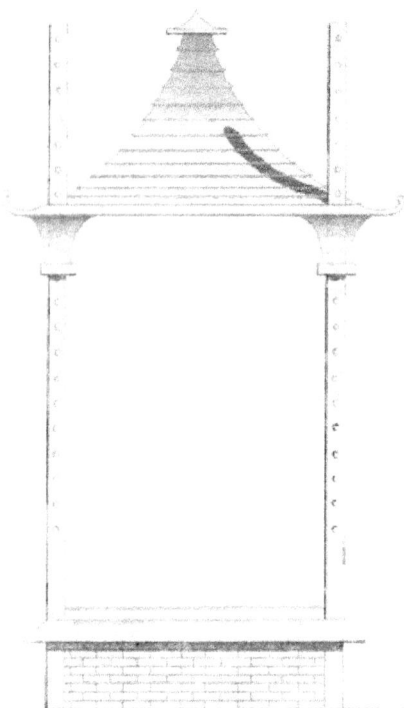 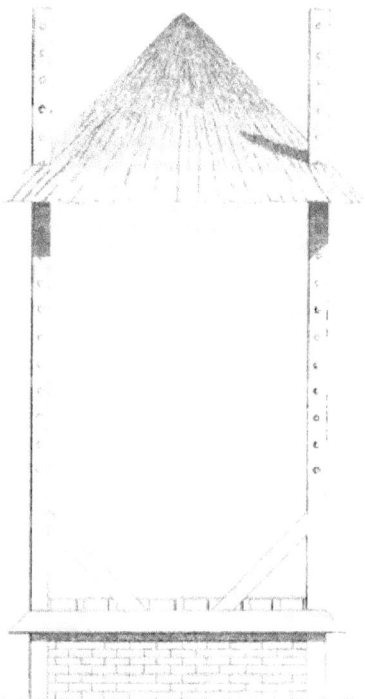

Barn with Board Covering.   Dutch Barn for Thatch.

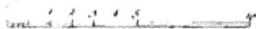

 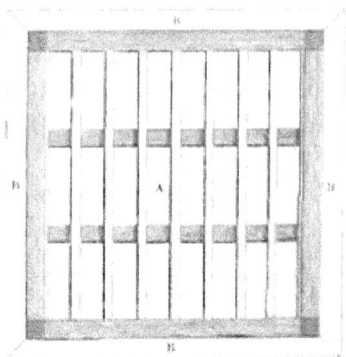

Curb or Plate of the Roof.

A. Plan of the Floor.
B. Board to prevent Vermin from
getting in the Hay or Corn.

Design for a Sheep Coat to
be built on a hill which seen
from a Genteel House forms
an agreeable Object.

a..Shepherds house.
b..Sheep-fold.
c.&c..Places for Hay.

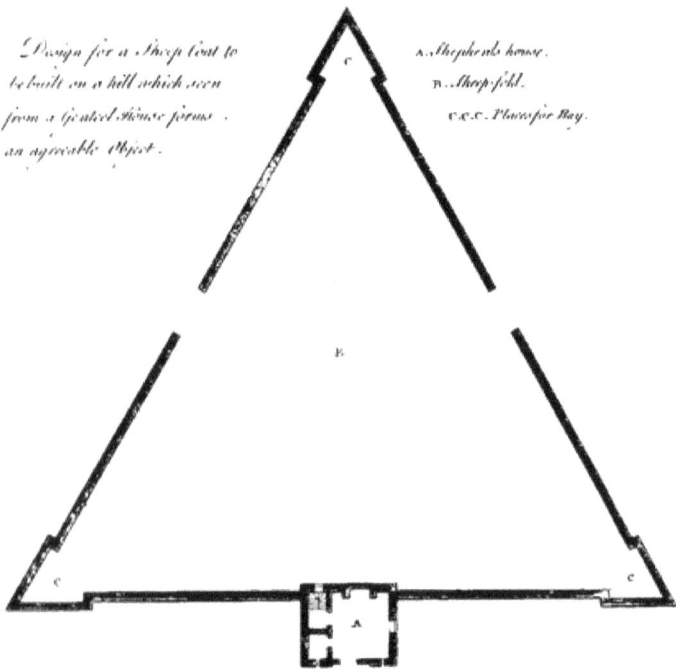

Facades to place before disagreable Objects.

A. Kitchin
B. Parlour
C. Milkhouse
D. Brewhouse
E. Wash house
F. Dairy
G. Barn
H. Stable or Cow house

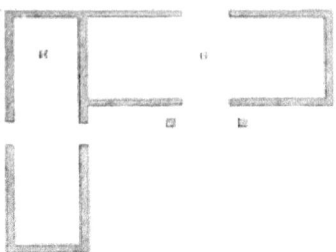

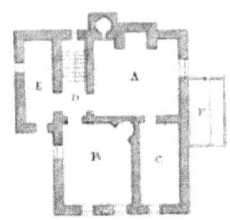

A. The Object supposed as shown by the scale line.

www.ingramcontent.com/pod-product-compliance
Lightning Source LLC
Chambersburg PA
CBHW022025190326
41519CB00010B/1599